TechnoAlchemy: Unleashing the Magic of Innovation

Introduction

Introducing "TechnoAlchemy: Unleashing the Magic of Innovation

In an era defined by rapid advancements and transformative breakthroughs, the pursuit of knowledge in technology and innovation has become paramount. This eBook serves as a comprehensive guide, unveiling the dynamic landscape of Pradyogiki (Technology) and Navachar (Innovation).

With a meticulous blend of scholarship and accessibility, this literary masterpiece delves deep into the heart of technological evolution and the art of pioneering change. We embark on a journey that transcends geographical boundaries, exploring the global tapestry of innovation from a British perspective.

The eBook is a treasure trove of insights, encompassing the marvels of artificial intelligence, blockchain, nanotechnology, and much more. It also scrutinizes the ethical dimensions of innovation, shedding light on its societal implications.

As we navigate through the digital age, the eBook equips readers with the tools and knowledge to not only adapt but also shape the future. It is a beacon for technophiles, inventors, and visionaries, providing the essential compass to navigate the uncharted waters of technology and innovation.

Join us on this enlightening odyssey as we uncover the boundless possibilities, challenges, and inspirations that lie at the intersection of Pradyogiki and Navachar. "Pradyogiki aur Navachar: The Ultimate Guide to Technology and Innovation" is your passport to a brighter, more informed future.

Index

14. Ethical Tech: Discuss the importance of ethical considerations in technological development and how to address emerging ethical challenges.

15. Innovation Ecosystems: Examine the role of innovation hubs, startups, and collaborative environments in driving technological progress.

CHAPTER 1

Digital Transformation: Navigating the Modern Age for Competitive Advantage

In the realm of modern business, the concept of "Digital Transformation" has emerged as a pivotal force reshaping how organizations operate, compete, and thrive. This seismic shift towards digitalization, often referred to as the Fourth Industrial Revolution, encapsulates a profound alteration in the way businesses conduct their affairs, interact with customers, and drive innovation. In this comprehensive exploration, we delve into the multifaceted dimensions of Digital Transformation, unravelling its essence, impact, and strategies, all through the lens of British English.

The Digital Imperative

In the face of a rapidly evolving technological landscape, businesses worldwide find themselves at a critical juncture. The digital age has ushered in an era where the ability to harness technology effectively is synonymous with survival. Traditional methods of conducting commerce are no longer sufficient, as consumers, markets, and competitors have all transitioned to digital platforms. The imperative to adapt and evolve is clear: digital transformation is no longer a choice but a necessity.

Understanding Digital Transformation

Digital Transformation extends far beyond the mere adoption of new technologies. It encompasses a holistic approach to integrating digital solutions into every facet of an organization's operations, culture, and customer engagement. At its core, it involves leveraging data, analytics, cloud computing, artificial intelligence, and other emerging technologies to drive innovation, streamline processes, enhance customer experiences, and ultimately gain a competitive edge.

The Pillars of Digital Transformation

transition and maximum efficiency gains. Cloud computing, IoT, and AI are among the key enablers of this integration.

5. Business Model Innovation: Organizations reevaluate and sometimes completely reinvent their business models to stay relevant. This may involve exploring new revenue streams, partnerships, or even transitioning from product-centric to service-centric models.

The Benefits of Digital Transformation

Embracing Digital Transformation brings forth a multitude of benefits:

Enhanced Efficiency: Automation and streamlined processes lead to increased operational efficiency, reducing costs and improving productivity.

Competitive Advantage: Organizations that successfully undergo Digital Transformation gain a significant edge in the marketplace. They are better equipped to respond to changing customer demands and market dynamics.

Improved Customer Experiences: Personalization and data-driven insights enable organizations to offer tailored experiences that resonate with customers, fostering loyalty and brand advocacy.

Innovation and Agility: Digital Transformation fosters a culture of innovation and adaptability, allowing businesses to swiftly respond to market shifts and emerging opportunities.

Data Security and Compliance: Robust cybersecurity measures are integral to Digital Transformation, ensuring the protection of sensitive data and compliance with regulations.

Challenges and Considerations

While Digital Transformation offers immense potential, it is not without its challenges. Organizations must navigate issues such as:

Resistance to Change: Employees may be resistant to new technologies and processes, requiring effective change management strategies.

Data Privacy and Security: The collection and use of data raise concerns about privacy and security. Organizations must address these issues to maintain trust.

Integration Complexities: Migrating from legacy systems to modern digital solutions can be complex and costly. It requires meticulous planning and execution.

Skills Gap: Organizations may face a shortage of talent with the necessary digital skills. Investing in training and upskilling is often essential.

The Future of Digital Transformation

As technology continues to advance at an unprecedented pace, the journey of Digital Transformation is far from over. It is an ongoing process, demanding perpetual adaptation and innovation. The organizations that will thrive in the modern age are those that not only embrace Digital Transformation but also make it an integral part of their DNA.

In the realm of British business, Digital Transformation is not merely a buzzword but a call to action. The UK has been at the forefront of technology and innovation for centuries, and this tradition continues in the digital era. British companies are leveraging their historical strengths in finance, creativity, and entrepreneurship to drive Digital Transformation and remain competitive on the global stage.

In conclusion, "Digital Transformation" is not a standalone event; it's an ongoing journey, a cultural shift, and a commitment to embracing the digital age with unwavering determination. In the pursuit of this transformation, organizations in the UK and around the world are poised to shape the future, drive innovation, and create

unparalleled value for their stakeholders and customers alike.

Chapter 2

Cybersecurity Mastery: Safeguarding You and Your Business in the Digital Age

In an increasingly digitized world, the critical importance of cybersecurity cannot be overstated. "Cybersecurity Mastery" represents the vanguard of our digital defense, arming individuals and businesses with the knowledge and strategies needed to protect against the ever-evolving landscape of cyber threats. This comprehensive exploration, steeped in British English, delves deep into the world of cybersecurity, unveiling its significance, challenges, and cutting-edge strategies.

The Digital Frontier and Its Vulnerabilities

The digital age has ushered in unprecedented connectivity, convenience, and innovation, but it has also opened the door to a myriad of cyber threats. Malicious actors, ranging from hackers to cybercriminal organizations, exploit vulnerabilities in our interconnected world, causing financial losses, data breaches, and damage to reputations. In the wake of such threats, cybersecurity emerges as a shield of paramount importance.

Understanding Cybersecurity

At its core, cybersecurity encompasses the practices, technologies, and strategies employed to safeguard digital systems, networks, and data from unauthorized access, damage, or theft. It is an intricate tapestry of measures designed to mitigate risks, protect assets, and maintain the confidentiality, integrity, and availability of information.

The Crucial Role of Education

In the realm of cybersecurity, knowledge is power. "Cybersecurity Mastery" serves as an educational beacon, illuminating the intricacies of this field. It empowers individuals and businesses alike with the ability to make informed decisions regarding their digital security. Awareness is the first line of defense, and this eBook equips readers with a comprehensive understanding of the cyber threat landscape.

Elements of Cybersecurity Mastery

1. Risk Assessment: Cybersecurity begins with a thorough assessment of potential vulnerabilities and threats. Identifying weaknesses in digital systems and networks is the foundation upon which protective strategies are built.

2. Threat Detection and Prevention: Cutting-edge tools and techniques are deployed to detect and prevent cyber threats in real-time. These include intrusion detection systems, firewalls, and advanced antivirus software.

3. Secure Password Practices: Passwords are the keys to digital assets. "Cybersecurity Mastery" emphasizes the importance of strong, unique passwords and advocates for the use of multi-factor authentication to enhance security.

4. Data Encryption: Data is often a prime target for cybercriminals. Encryption transforms data into unreadable code, protecting it from unauthorized access.

5. Employee Training: A cybersecurity strategy is only as strong as its weakest link. Training employees to recognize and respond to phishing attacks and other threats is vital.

6. Incident Response and Recovery: Despite best efforts, breaches can occur. "Cybersecurity Mastery" guides readers through the steps of incident response and recovery, minimizing the impact of cyber incidents.

7. Compliance and Regulations: The eBook sheds light on the importance of adhering to cybersecurity regulations and standards, especially in industries where sensitive data is handled.

The Human Factor

One of the most significant challenges in cybersecurity lies in human behavior. Cybercriminals often exploit human errors, using social engineering tactics to gain access to systems. "Cybersecurity Mastery" underscores the importance of cultivating a cybersecurity-conscious culture within organizations and educating individuals to recognize and resist manipulation attempts.

The Business Imperative

For businesses, the stakes are higher than ever. Cyberattacks can result in crippling financial losses, reputational damage, and legal repercussions. "Cybersecurity Mastery" provides businesses with the guidance to implement robust cybersecurity strategies, protect their digital assets, and safeguard their future.

The Evolving Landscape

As technology continues to advance, so do cyber threats. The eBook emphasizes the need for continuous vigilance and adaptation in the face of new threats, including ransomware, zero-day vulnerabilities, and emerging attack vectors.

The British Perspective

The United Kingdom has long been a global hub for technology and innovation. British businesses have a unique opportunity to lead the way in cybersecurity, leveraging their expertise in technology and their commitment to excellence. The eBook acknowledges the contributions of British organizations in the field and encourages them to champion cybersecurity best practices.

Conclusion

In conclusion, "Cybersecurity Mastery" is not merely a book; it is a call to action. It recognizes that in our digitally connected world, the battle for cybersecurity is ongoing and dynamic. With knowledge as their weapon, individuals and businesses can fortify their defenses, thwart cyber threats, and navigate the digital age with confidence. This eBook serves as a testament to the power of education, vigilance, and a commitment to safeguarding the digital realm. In the face of evolving cyber threats, "Cybersecurity Mastery" equips readers to protect themselves, their businesses, and the digital future

CHAPTER 3

Sustainable Solutions: Pioneering a Greener Future through Innovation

In an era defined by environmental challenges, the pursuit of sustainable solutions has become a global imperative. "Sustainable Solutions" serves as a compass, guiding individuals and businesses towards innovative technologies that are shaping a more eco-friendly and sustainable future. This comprehensive exploration, steeped in British English, unveils the essence of sustainability, highlights pressing environmental issues, and showcases cutting-edge technologies that hold the promise of a greener tomorrow.

The Urgency of Sustainability

The twenty-first century has witnessed an unprecedented awareness of the environmental crises facing our planet. Climate change, resource depletion, pollution, and biodiversity loss have become pressing concerns that demand immediate attention. At the heart of this global challenge lies the concept of sustainability—an ethos that seeks to balance the needs of the present without compromising the ability of future generations to meet their own needs.

Understanding Sustainability

Sustainability is not merely a buzzword; it's a way of life, a commitment to responsible stewardship of our planet's resources. At its core, sustainability advocates for practices that reduce harm to the environment, conserve resources, promote economic and social well-being, and foster resilience in the face of global challenges.

The Technological Leap

In the quest for sustainability, technology emerges as a potent ally. Innovative solutions driven by scientific advancements and creative thinking are redefining how we interact with the environment. "Sustainable Solutions" bridges the gap between environmental consciousness and technological prowess, showcasing how innovative technologies are spearheading the sustainability revolution.

Key Sustainability Challenges

"Sustainable Solutions" begins by illuminating the primary environmental challenges that humanity faces:

*1. Climate Change: The eBook underscores the urgency of mitigating climate change, highlighting the consequences of rising temperatures, extreme weather events, and the melting of polar ice caps.

*2. Resource Depletion: It explores the unsustainable exploitation of finite resources such as fossil fuels, minerals, and freshwater.

*3. Biodiversity Loss: The eBook emphasizes the importance of preserving biodiversity, which is essential for ecosystem health and human well-being.

*4. Waste Management: It delves into the mounting problem of waste generation and the importance of adopting circular economy principles.

Innovative Sustainable Technologies

The eBook then delves into a myriad of innovative technologies that are propelling the sustainability agenda:

*1. Renewable Energy: Wind, solar, and hydropower are transforming the energy landscape, reducing carbon emissions and dependence on fossil fuels.

*2. Electric Vehicles: The transition to electric vehicles (EVs) is revolutionizing the transportation sector, reducing air pollution and greenhouse gas emissions.

*3. Circular Economy: "Sustainable Solutions" explores the concept of a circular economy, where products and materials are reused, refurbished, or recycled to minimize waste.

*4. Green Building Design: The eBook showcases green building practices that enhance energy efficiency and reduce the environmental impact of construction.

*5. Waste-to-Energy: It discusses technologies that convert waste into energy, reducing landfill usage and harnessing the energy potential of organic matter.

*6. Precision Agriculture: The eBook highlights precision agriculture techniques that optimize resource use, increase crop yields, and minimize environmental impact.

*7. Carbon Capture and Storage: It explores carbon capture technologies that capture and store carbon dioxide emissions, helping to combat climate change.

The British Perspective

The United Kingdom, with its rich history of innovation and environmental stewardship, plays a pivotal role in the global sustainability movement. British businesses and institutions are at the forefront of developing and implementing sustainable technologies. "Sustainable Solutions" acknowledges the contributions of the UK and encourages the nation to continue leading by example.

A Call to Action

In conclusion, "Sustainable Solutions" is not just a book; it is a call to action. It calls upon individuals, businesses, and nations to recognize their role in safeguarding the planet for future generations. It champions the idea that sustainability is not a sacrifice but an opportunity—a chance to innovate, prosper, and ensure the well-being of all living creatures.

As we navigate the complex challenges of the modern world, "Sustainable Solutions" reminds us that the power to change lies within our grasp. By embracing innovative technologies and sustainable practices, we can embark on a path towards a greener, more prosperous future—one where the delicate balance between human needs and environmental preservation is upheld. The eBook serves as a testament to the power of ingenuity, cooperation, and the enduring human spirit that seeks to leave a positive mark on this planet we call home.

CHAPTER 4

AI Augmentation: Elevating Human Potential Across Industries

In the digital age, the integration of Artificial Intelligence (AI) into various facets of our lives has transformed the way we work, innovate, and thrive. "AI Augmentation" stands as a beacon, illuminating the profound ways in which AI is enhancing human capabilities across diverse industries. In this comprehensive exploration, steeped in British English, we delve into the essence of AI augmentation, its transformative impact, and its role in shaping a future where human ingenuity and machine intelligence intertwine.

The AI Augmentation Paradigm

AI augmentation represents a shift from the traditional fear of machines replacing humans to a more harmonious synergy where AI enhances human abilities. This paradigm recognizes that AI technologies, such as machine learning, natural language processing, and computer vision, are tools that amplify human potential, creativity, and productivity.

Understanding AI Augmentation

At its core, AI augmentation harnesses the power of artificial intelligence to augment, rather than replace, human skills and expertise. It leverages AI to automate routine tasks, analyze vast datasets, and make predictions or recommendations, enabling humans to focus on higher-order thinking, creativity, and problem-solving.

AI Across Industries

"AI Augmentation" unveils the versatile applications of AI in diverse sectors:

*1. Healthcare: In the healthcare sector, AI augments medical professionals' diagnostic accuracy, streamlines administrative tasks, and accelerates drug discovery, ultimately improving patient care.

*2. Finance: AI-driven algorithms enhance financial decision-making by analyzing market trends, managing portfolios, and identifying potential fraud.

*3. Manufacturing: AI-powered robots and automation systems optimize production processes, reduce errors, and enhance quality control.

*4. Education: AI aids educators in personalizing learning experiences, automating administrative tasks, and providing data-driven insights to enhance teaching methods.

*5. Retail: AI-driven recommendation engines and chatbots improve customer experiences and assist in inventory management.

*6. Transportation: AI augments transportation systems with autonomous vehicles, traffic management, and predictive maintenance.

*7. Creativity: AI generates art, music, and literature, pushing the boundaries of human creativity.

The British Perspective

The United Kingdom, with its rich history of innovation and technological advancements, has been at the forefront of AI research and application. British businesses and academic institutions play a pivotal role in shaping the AI landscape, emphasizing responsible AI development and ethical considerations. "AI Augmentation" acknowledges the contributions of the UK to AI progress and encourages the nation to continue leading in this transformative field.

AI Ethics and Responsible Augmentation

While AI augmentation brings immense benefits, it also raises ethical questions and concerns. "AI Augmentation" emphasizes the importance of responsible AI development, including transparency, fairness, and accountability. It underscores the need for regulations and guidelines to ensure that AI augments human potential without compromising ethical principles.

AI and the Future of Work

The eBook addresses the evolving nature of work in the age of AI augmentation. It recognizes that as AI takes over routine tasks, humans are liberated to focus on tasks that require emotional intelligence, creativity, critical thinking, and complex problem-solving—skills that are inherently human and difficult for AI to replicate.

AI and Challenges Ahead

While AI augmentation holds immense promise, it is not without challenges. These include concerns about bias in AI algorithms, data privacy, and the need for continuous upskilling to adapt to the evolving job landscape. "AI Augmentation" encourages individuals and organizations to be proactive in addressing these challenges.

A Collaborative Future

In conclusion, "AI Augmentation" serves as a reminder that the future is not a battleground between humans and machines but a collaborative frontier where AI amplifies human potential. It calls upon individuals, businesses, and nations to embrace AI augmentation as a powerful tool for progress and innovation. The eBook underscores that our ability to harness AI for the betterment of society lies in our capacity to use it responsibly, ethically, and creatively.

As we stand at the intersection of human ingenuity and artificial intelligence, "AI Augmentation" is a testament to the boundless possibilities that arise when we combine our innate human qualities with the capabilities of AI. It is a roadmap to a future where the synergy between humans and machines empowers us to overcome challenges, spark innovation, and achieve new heights of progress across industries, ultimately creating a brighter and more prosperous world for all.

CHAPTER 5

Blockchain Revolution: Unveiling the Potential of a Transformative Technology

In the rapidly evolving landscape of technology, few innovations have sparked as much excitement and disruption as blockchain. "Blockchain Revolution" serves as a guide, illuminating the immense potential of blockchain technology, its applications ranging from cryptocurrencies to supply chain management, and its profound impact on industries and society at large. In this comprehensive exploration, steeped in British English, we delve into the essence of blockchain, its transformative capabilities, and the promising future it heralds.

The Blockchain Paradigm Shift

Blockchain, often dubbed as the "Internet of Value," represents a fundamental shift in how we record, share, and trust information. It introduces the concept of decentralized and immutable ledgers, where transactions are transparent, secure, and tamper-proof. This paradigm shift challenges traditional centralized systems and intermediaries, offering a more transparent and trustless alternative.

Understanding Blockchain

At its core, blockchain is a distributed ledger technology that records transactions across a network of computers, known as nodes. These transactions are grouped into blocks, cryptographically linked together in chronological order, forming a chain. The decentralized nature of blockchain ensures that no single entity controls the entire network, enhancing security and trust.

Cryptocurrencies: A Trailblazing Application

Cryptocurrencies, such as Bitcoin, serve as the pioneering application of blockchain technology. "Blockchain Revolution" unravels the significance of cryptocurrencies, emphasizing their role as digital assets that enable secure, borderless, and pseudonymous transactions. Bitcoin, in particular, has emerged as a decentralized digital currency, disrupting traditional financial systems and challenging the concept of central banks.

Beyond Cryptocurrencies: Real-World Applications

While cryptocurrencies garner much attention, blockchain's potential extends far beyond digital currencies:

Supply Chain Management: "Blockchain Revolution" highlights how blockchain enhances transparency and traceability in supply chains. It enables the tracking of goods from origin to destination, reducing fraud, ensuring product authenticity, and enhancing the efficiency of logistics.

Smart Contracts: Smart contracts, self-executing agreements with predefined rules, are explored as a powerful blockchain application. These contracts automate processes, reducing the need for intermediaries, and are used in fields ranging from legal agreements to insurance.

Voting Systems: The eBook underscores how blockchain can revolutionize voting systems, ensuring secure and verifiable elections, and promoting civic participation.

Healthcare: Blockchain's potential in healthcare is revealed, from securely managing patient data to tracking the authenticity of pharmaceuticals.

Intellectual Property: Blockchain can be used to timestamp and authenticate intellectual property rights, safeguarding the interests of creators.

The British Perspective

The United Kingdom, renowned for its financial sector and technological innovation, plays a pivotal role in the global blockchain revolution. British businesses, startups, and academic institutions are at the forefront of blockchain research and application. "Blockchain Revolution" acknowledges the UK's contributions and encourages continued leadership in the blockchain space.

Challenges and Ethical Considerations

While blockchain offers transformative potential, it is not without challenges. The eBook addresses issues like scalability, energy consumption (particularly in proof-of-work blockchains like Bitcoin), regulatory concerns, and the need for ethical governance. It emphasizes the importance of responsible blockchain development and collaboration between industry, government, and academia to address these challenges.

A Transformative Future

In conclusion, "Blockchain Revolution" is not just a book; it's a roadmap to a transformative future. It underscores that blockchain is not a mere technological innovation but a catalyst for societal change. As blockchain technology continues to mature, it has the potential to reshape industries, empower individuals, and promote transparency and trust across various sectors.

The eBook serves as a reminder that the blockchain revolution is not about overthrowing established systems but about building a more inclusive, secure, and efficient world. It calls upon individuals, businesses, and governments to embrace blockchain as a powerful tool for innovation and to navigate the evolving landscape with ethics, responsibility, and a commitment to positive change.

As we stand at the threshold of a blockchain-powered future, "Blockchain Revolution" inspires us to reimagine systems, reinvent processes, and reimpose trust in a world driven by decentralization and transparency. It is a testament to the boundless possibilities that arise when technology aligns with values, paving the way for a brighter, more equitable future for all.

CHAPTER 6

Smart Cities: A Glimpse into the Urban Evolution of Tomorrow

In an increasingly urbanized world, the concept of "Smart Cities" has emerged as a transformative force, reshaping the way we live, work, and interact with urban environments. "Smart Cities" stands as a visionary guide, shedding light on the remarkable evolution of urban areas into intelligent, interconnected hubs where technology enhances efficiency, sustainability, and the quality of life. This comprehensive exploration, steeped in British English, unveils the essence of smart cities, their profound impact, and their role in shaping the future of urban living.

The Urbanization Challenge

Urbanization is an undeniable global trend, with the majority of the world's population now residing in cities. While cities offer economic opportunities and cultural vibrancy, rapid urbanization presents a host of challenges, including traffic congestion, environmental degradation, inadequate infrastructure, and strains on resources.

Understanding Smart Cities

At its core, a smart city is a place where technology, data, and innovation converge to optimize urban living. Smart cities leverage digital solutions to enhance infrastructure, services, and the overall quality of life for residents. They prioritize sustainability, efficiency, and inclusivity while fostering economic growth and innovation.

The Four Pillars of Smart Cities

"Smart Cities" unveils the four key pillars that underpin the smart city concept:

Infrastructure: The eBook explores how advanced infrastructure, including high-speed broadband, IoT sensors, and 5G connectivity, forms the backbone of smart cities, enabling seamless data exchange and communication.

Data and Technology: It emphasizes the role of data analytics, AI, and machine learning in harnessing the vast amounts of data generated in cities to inform decision-making, optimize resource allocation, and improve services.

Sustainability: Smart cities prioritize sustainability by implementing eco-friendly practices such as renewable energy generation, waste reduction, efficient transportation systems, and green building designs.

Citizen Engagement: The eBook highlights the importance of engaging citizens in the planning and decision-making processes. It explores initiatives that encourage citizen participation, transparency, and feedback.

Applications of Smart Cities

"Smart Cities" delves into the multifaceted applications of smart city technologies:

Transportation: The eBook discusses how smart transportation systems, including real-time traffic management, autonomous vehicles, and efficient public transit, reduce congestion and enhance mobility.

Energy: It explores how smart grids, renewable energy sources, and energy-efficient technologies promote sustainability and resilience in the face of climate change.

Healthcare: Smart healthcare solutions, such as telemedicine and remote monitoring, improve access to healthcare services and enhance public health outcomes.

Security: The eBook highlights the role of advanced surveillance systems, IoT-enabled devices, and predictive analytics in enhancing urban safety and emergency response.

E-Governance: It emphasizes how digital governance platforms streamline administrative processes, improve service delivery, and promote transparency.

The British Perspective

The United Kingdom, with its vibrant cities and a strong tradition of innovation, is actively contributing to the smart city movement. British cities are at the forefront of adopting smart technologies to enhance urban living and address contemporary challenges. "Smart Cities" acknowledges the UK's role in shaping the future of urbanization and encourages continued leadership in smart city initiatives.

Challenges and Considerations

While smart cities offer tremendous promise, they also face challenges related to data privacy, security, digital inclusion, and ethical concerns. "Smart Cities" underscores the importance of addressing these challenges through responsible governance, robust cybersecurity measures, and inclusive strategies to ensure that the benefits of smart cities are accessible to all residents.

A Vision for the Future

In conclusion, "Smart Cities" is not just a book; it is a vision for the future of urban living. It underscores that the smart city revolution is not about technology for its own sake but about improving the lives of city dwellers, fostering sustainability, and promoting economic growth.

The eBook serves as a reminder that the cities of tomorrow can be intelligent, responsive, and inclusive. It calls upon governments, businesses, and citizens to collaborate in building cities that prioritize efficiency, sustainability, and the well-being of all residents. As we stand at the precipice of urban transformation, "Smart Cities" inspires us to reimagine our cities as vibrant, interconnected ecosystems where technology serves as a catalyst for progress, innovation, and a higher quality of life for all. It is a testament to the potential of human ingenuity in creating cities that are not just smart but also truly livable.

CHAPTER 7

The Internet of Things (IoT): Unraveling the Interconnected Tapestry of Modern Life

In our increasingly digitized world, the Internet of Things (IoT) stands as a transformative force, interweaving the fabric of our daily lives with a network of interconnected devices. "The Internet of Things (IoT)" serves as an illuminating guide, offering a comprehensive understanding of this phenomenon, its myriad applications, and its profound impact on how we live, work, and interact with technology. In this exploration, steeped in British English, we delve into the essence of IoT, its transformative potential, and its role in shaping the future of daily life.

The IoT Paradigm

The Internet of Things represents a paradigm shift in our relationship with technology. It extends the internet's reach beyond computers and smartphones to a vast array of everyday objects. In essence, IoT refers to a network of interconnected devices—ranging from household appliances to industrial machinery—that collect, share, and act upon data.

Understanding IoT

At its core, IoT revolves around three fundamental components:

Devices and Sensors: IoT devices are equipped with sensors, actuators, and communication modules that enable them to collect data and interact with their environment. These devices can range from smart thermostats and fitness trackers to industrial sensors on factory machinery.

Connectivity: IoT devices are connected to the internet or private networks, allowing them to transmit data to centralized systems or other devices. This connectivity can be wired (e.g., Ethernet) or wireless (e.g., Wi-Fi, cellular, or Low-Power Wide-Area Networks).

Data Analytics and Action: The data collected by IoT devices is processed, analyzed, and often used to trigger actions or inform decision-making. Advanced analytics, artificial intelligence (AI), and machine learning play a crucial role in extracting insights from IoT-generated data.

The Impact on Daily Life

"The Internet of Things (IoT)" offers a glimpse into how IoT technologies are reshaping various aspects of our daily lives:

Smart Homes: IoT-enabled devices, such as smart thermostats, lights, and voice-controlled assistants, are enhancing convenience, energy efficiency, and security in our homes.

Health and Wellness: Wearable IoT devices, like fitness trackers and health monitors, are empowering individuals to take charge of their health by providing real-time data and insights.

Transportation: IoT sensors in vehicles and infrastructure are optimizing traffic flow, reducing congestion, and improving road safety. Moreover, the rise of autonomous vehicles is poised to revolutionize personal and public transportation.

Agriculture: IoT sensors and data analytics are revolutionizing agriculture, enabling farmers to monitor soil conditions, crop health, and weather patterns, leading to more efficient and sustainable farming practices.

Retail: IoT technologies are enhancing the customer experience through personalized marketing, inventory management, and cashier-free stores.

Industrial IoT (IIoT): In the industrial sector, IIoT is improving operational efficiency, predictive maintenance, and asset utilization in manufacturing, energy, and logistics.

Environmental Monitoring: IoT sensors are used to monitor air and water quality, track wildlife, and detect natural disasters, aiding in environmental conservation and disaster response.

The British Perspective

The United Kingdom has been at the forefront of IoT innovation, with British companies and institutions actively contributing to its development and deployment. "The Internet of Things (IoT)" recognizes the UK's role in shaping the IoT landscape and encourages continued leadership in IoT-related research, development, and ethical considerations.

Challenges and Considerations

While IoT offers immense potential, it also presents challenges and considerations, including privacy concerns, data security, interoperability issues, and the ethical use of IoT data. The eBook underscores the importance of addressing these challenges through responsible IoT design, robust security measures, and regulatory frameworks.

A Vision for the Future

In conclusion, "The Internet of Things (IoT)" is not just a book; it is a window into a future where the physical and digital worlds converge seamlessly. It underscores that IoT is not simply about connecting devices but about transforming the way we live and interact with our surroundings.

The eBook serves as a reminder that the IoT revolution is not about technology for its own sake but about improving our quality of life, enhancing sustainability, and solving real-world problems. It calls upon individuals, businesses, and governments to embrace IoT as a powerful tool for innovation and to navigate the evolving landscape with ethics, responsibility, and a commitment to positive change.

As we stand at the threshold of a hyperconnected world, "The Internet of Things (IoT)" inspires us to reimagine our daily lives as a tapestry of interconnected devices working harmoniously to make our lives more convenient, efficient, and sustainable. It is a testament to the potential of human ingenuity in creating a future where technology serves as a catalyst for progress and where IoT enriches the human experience in ways we are only beginning to imagine.

CHAPTER 8

Biotechnology Breakthroughs: Pioneering Discoveries Shaping Our Future

In the ever-evolving realm of science and innovation, the field of biotechnology stands as a beacon of hope and progress, driving breakthroughs that have the potential to revolutionize our understanding of life itself. "Biotechnology Breakthroughs" serves as an illuminating guide, offering a comprehensive insight into the cutting-edge developments within this field. From gene editing to personalized medicine, this exploration, steeped in British English, delves into the essence of biotechnology, its transformative impact, and its role in shaping the future of healthcare, agriculture, and beyond.

The Biotechnology Frontier

Biotechnology represents a dynamic convergence of biology, genetics, and technology, unleashing the power of scientific discovery to address some of humanity's most pressing challenges. At its core, biotechnology harnesses the fundamental processes of living organisms to develop innovative solutions, with applications spanning healthcare, agriculture, and environmental sustainability.

Understanding Biotechnology

Fundamentally, biotechnology encompasses the manipulation of biological systems and the utilization of cellular and biomolecular processes to create products and technologies that improve our quality of life. It encompasses a vast array of techniques, from genetic engineering to bioprocessing, all with the common goal of harnessing the potential of life's building blocks.

Pioneering Biotechnological Breakthroughs

"Biotechnology Breakthroughs" sheds light on a selection of groundbreaking developments within the field:

Gene Editing: The eBook explores the revolutionary CRISPR-Cas9 technology, which enables precise modification of DNA, opening new avenues for treating genetic diseases and advancing genetic research.

Personalized Medicine: It discusses how biotechnology is reshaping healthcare through personalized treatment plans, tailored to an individual's genetic makeup and medical history, thus maximizing effectiveness and minimizing side effects.

Biopharmaceuticals: The eBook highlights the role of biotechnology in the production of therapeutic proteins, monoclonal antibodies, and vaccines, with applications ranging from cancer treatment to combating infectious diseases.

Agricultural Biotechnology: It delves into the development of genetically modified organisms (GMOs) and the potential of biotechnology to enhance crop yields, improve nutritional content, and reduce the environmental impact of agriculture.

Stem Cell Research: The eBook explores the promising applications of stem cells in regenerative medicine, tissue engineering, and the treatment of degenerative diseases.

Synthetic Biology: It discusses the emerging field of synthetic biology, where biological parts and systems are designed and engineered for diverse applications, from biofuel production to bioremediation.

The British Perspective

The United Kingdom has a rich history of scientific innovation, and British researchers and institutions have played a significant role in advancing biotechnology. "Biotechnology Breakthroughs" recognizes the UK's contributions to biotechnological research and encourages continued leadership in ethical biotechnological development.

Challenges and Ethical Considerations

While biotechnology offers immense promise, it also raises ethical, regulatory, and safety considerations. The eBook emphasizes the importance of ethical conduct in research, rigorous safety protocols, and transparent regulation to ensure that biotechnological advancements benefit society responsibly.

A Vision for the Future

In conclusion, "Biotechnology Breakthroughs" is not just a book; it is a window into a future where science and technology are redefining the boundaries of what is possible. It underscores that biotechnology is not just about manipulating genes but about alleviating suffering, improving human health, enhancing food security, and preserving the environment.

The eBook serves as a reminder that the biotechnology revolution is not about scientific achievements for their own sake but about creating a better, healthier, and more sustainable world. It calls upon individuals, businesses, and governments to embrace biotechnology as a powerful tool for innovation, to navigate the evolving landscape with ethics and responsibility, and to channel the boundless potential of biotechnology toward addressing the most pressing challenges of our time.

As we stand at the threshold of a biotechnological future, "Biotechnology Breakthroughs" inspires us to reimagine the possibilities of life sciences, celebrate human ingenuity, and continue the journey of discovery, innovation, and collaboration to shape a brighter future for all.

CHAPTER 9

Space Exploration: Navigating the Cosmos on a Quest for Discovery

In the annals of human history, few endeavors have captured the imagination and aspirations of humanity quite like space exploration. "Space Exploration" serves as an enthralling guide, chronicling the latest advancements in space technology and humanity's relentless race to explore new frontiers beyond our terrestrial boundaries. In this comprehensive exploration, steeped in British English, we delve into the essence of space exploration, its transformative impact, and its role in shaping the future of science, technology, and our understanding of the cosmos.

The Cosmic Quest

Space exploration embodies humanity's innate curiosity, our ceaseless quest to unravel the mysteries of the universe, and our unrelenting desire to push the boundaries of what is possible. It represents a journey into the unknown, a journey that not only advances our knowledge but also holds the promise of profound discoveries that could change the course of history.

Understanding Space Exploration

At its core, space exploration encompasses a range of scientific, technological, and human endeavors aimed at investigating and venturing beyond Earth's atmosphere. It spans robotic missions, space telescopes, planetary exploration, space station operations, and human spaceflight. These endeavors are driven by a thirst for knowledge, a quest for scientific discovery, and the exploration of resources and habitats beyond our planet.

Recent Advancements and Missions

"Space Exploration" delves into some of the most recent advancements and missions that have captivated the world's attention:

Mars Exploration: The eBook explores the Mars Perseverance Rover mission, which seeks to unravel the Red Planet's geological history and search for signs of past microbial life.

International Space Station (ISS): It discusses the role of the ISS as a space laboratory, fostering international collaboration in science and technology research.

Private Space Endeavors: The eBook highlights the growing role of private companies like SpaceX, Blue Origin, and Virgin Galactic in advancing space exploration and commercial space travel.

Astronomy and Space Telescopes: It explores the monumental contributions of space telescopes like the Hubble Space Telescope and the James Webb Space Telescope to our understanding of the cosmos.

Moon Missions: The eBook covers recent lunar missions, including China's Chang'e missions and NASA's Artemis program, which aims to return humans to the Moon and establish a sustainable lunar presence.

The British Perspective

The United Kingdom has a storied history of contributions to space exploration, from early rocketry experiments to satellite technology and planetary science missions. "Space Exploration" acknowledges the UK's role in the global space community and encourages continued leadership in space research and technology development.

Challenges and Aspirations

While space exploration holds immense promise, it also poses numerous challenges. These include the harsh conditions of space, the need for sustainable space habitats, mitigating space debris, and ensuring the responsible use of celestial resources. The eBook emphasizes the importance of international collaboration, space governance, and sustainable practices in addressing these challenges.

A Vision for the Future

In conclusion, "Space Exploration" is not just a book; it is a testament to human ingenuity, curiosity, and our unyielding spirit of exploration. It underscores that space exploration is not merely a scientific endeavor but a journey of human discovery and progress.

The eBook serves as a reminder that space exploration is a shared human endeavor, one that transcends national boundaries and political differences. It calls upon individuals, nations, and the global community to unite in the pursuit of knowledge, scientific discovery, and the peaceful exploration of the cosmos.

As we stand at the threshold of a new era in space exploration, "Space Exploration" inspires us to gaze skyward, to reach for the stars, and to continue the age-old tradition of exploration and discovery that defines our species. It is a roadmap to a future where humanity's journey into the cosmos continues to inspire, educate, and illuminate the mysteries of the universe for generations to come.

CHAPTER 10

an increasingly digital world, the dawn of 5G technology stands as a transformative moment, promising lightning-fast connectivity that has the potential to reshape how we communicate, work, and interact with technology. "5G Connectivity" serves as an enlightening guide, providing a comprehensive understanding of 5G networks, their capabilities, and the profound implications they hold for communication and connectivity. In this exploration, steeped in British English, we delve into the essence of 5G technology, its transformative impact, and its role in shaping the future of connectivity.

The 5G Revolution

The advent of 5G technology represents a quantum leap in telecommunications, offering unparalleled speed, low latency, and network reliability. It marks a significant advancement from previous generations of wireless technology, promising not only faster internet speeds but also a vast array of applications that will transform industries and enhance our daily lives.

Understanding 5G

At its core, 5G is the fifth generation of wireless technology, succeeding 4G (LTE). It leverages advanced technologies and frequencies to deliver significantly faster data speeds, reduced latency (the time it takes for data to travel), and the ability to connect a vast number of devices simultaneously. These improvements are poised to revolutionize various sectors, from telecommunications to healthcare and transportation.

Key Features and Capabilities

"5G Connectivity" explores the key features and capabilities that define this groundbreaking technology:

High-Speed Connectivity: 5G networks offer data speeds that are up to 100 times faster than 4G, enabling near-instant downloads, seamless streaming, and lag-free online experiences.

Low Latency: With latency as low as one millisecond, 5G minimizes delays in data transmission, making it ideal for applications like augmented reality (AR), virtual reality (VR), and autonomous vehicles.

Massive IoT Connectivity: 5G can support a massive number of Internet of Things (IoT) devices, facilitating the growth of smart cities, smart homes, and industrial automation.

Enhanced Mobile Broadband (eMBB): The eBook discusses how 5G enhances mobile broadband, providing a reliable and high-speed internet experience, even in densely populated areas.

Network Slicing: 5G introduces the concept of network slicing, allowing the customization of network resources for specific applications, such as telemedicine, gaming, or autonomous driving.

Applications and Implications

"5G Connectivity" delves into the myriad of applications and implications that 5G networks offer:

Telecommunications: It highlights the potential for high-definition video calls, seamless streaming, and improved network coverage, even in remote areas.

Healthcare: 5G facilitates telemedicine, remote patient monitoring, and the rapid transfer of medical data, revolutionizing healthcare delivery.

Autonomous Vehicles: The eBook explores how 5G enables real-time communication between vehicles, traffic infrastructure, and pedestrians, making autonomous driving safer and more efficient.

Smart Cities: 5G plays a pivotal role in the development of smart cities, with applications ranging from efficient public transportation to smart energy grids and waste management.

Entertainment and Gaming: It discusses how 5G enhances gaming experiences with low-latency multiplayer gaming and AR/VR applications.

Industry 4.0: In the industrial sector, 5G enables advanced robotics, remote monitoring, and predictive maintenance, leading to increased efficiency and productivity.

The British Perspective

The United Kingdom has been actively engaged in the deployment and development of 5G technology. British companies and research institutions are contributing to the global advancement of 5G networks and applications. "5G Connectivity" acknowledges the UK's role in shaping the future of connectivity and encourages continued leadership in this field.

Challenges and Considerations

While 5G brings significant benefits, it also raises challenges related to network security, privacy, and the digital divide. The eBook emphasizes the importance of addressing these challenges through robust cybersecurity measures, responsible data handling, and efforts to bridge the digital divide to ensure equitable access to 5G's benefits.

A Vision for the Future

In conclusion, "5G Connectivity" is not just a book; it is a glimpse into a future where connectivity knows no bounds. It underscores that 5G is not merely an upgrade to our current networks but a gateway to a new era of innovation, connectivity, and economic growth.

The eBook serves as a reminder that 5G technology has the potential to transform industries, enhance our daily lives, and bridge geographical divides. It calls upon individuals, businesses, and governments to embrace 5G as a catalyst for progress, to harness its capabilities for the greater good, and to navigate the evolving landscape with ethics, responsibility, and a commitment to equitable access for all.

As we stand at the precipice of a 5G-powered future, "5G Connectivity" inspires us to imagine a world where connectivity is seamless, communication is instantaneous, and innovation knows no limits. It is a testament to the boundless possibilities that arise when technology aligns with human ingenuity, bringing us closer to a more connected and prosperous future for all.

CHAPTER 11

Robotics and Automation: Revolutionizing Industries with Cutting-Edge Technology

In the ever-evolving landscape of technology, robotics and automation have emerged as transformative forces, reshaping industries and redefining the way we work, manufacture, and deliver services. "Robotics and Automation" serves as a comprehensive guide, unveiling the remarkable impact of these technologies across various sectors, from manufacturing to healthcare and logistics. In this exploration, steeped in British English, we delve into the essence of robotics and automation, their disruptive potential, and their pivotal role in shaping the future of industries.

The Robotics and Automation Revolution

Robotics and automation represent a paradigm shift in how we approach tasks and processes. They encompass a broad spectrum of technologies, from autonomous robots on factory floors to intelligent software systems that streamline administrative tasks, all with the common goal of increasing efficiency, accuracy, and productivity.

Understanding Robotics and Automation

At its core, robotics involves the design, construction, and operation of robots—mechanical or virtual entities capable of performing tasks with varying degrees of autonomy. Automation, on the other hand, refers to the use of technology and software to perform tasks with minimal human intervention.

Key Components and Technologies

"Robotics and Automation" explores the key components and technologies that drive this transformation:

Robotics Hardware: It discusses the various types of robots, including industrial robots, service robots, and collaborative robots (cobots), and their applications in industries.

Artificial Intelligence (AI): The eBook highlights how AI and machine learning enable robots and automation systems to adapt, learn, and make decisions based on data and feedback.

Sensors: Sensors play a crucial role in automation by providing robots and machines with real-time data about their surroundings, facilitating precise and safe operations.

Internet of Things (IoT): The eBook delves into how IoT connects devices and machines, enabling them to communicate and share data for improved automation and efficiency.

Applications Across Industries

"Robotics and Automation" explores how these technologies are revolutionizing diverse sectors:

Manufacturing: It showcases how robots and automation streamline production lines, reduce errors, and enhance quality control in manufacturing processes.

Healthcare: The eBook discusses the role of robotic surgery, automated drug dispensing, and telemedicine in improving patient care and efficiency in healthcare settings.

Logistics and Warehousing: It highlights how automation and autonomous robots are reshaping supply chains, from automated warehouses to delivery drones.

Agriculture: Automation is transforming farming with autonomous tractors, drones for crop monitoring, and robotic harvesters.

Construction: Robots are used in construction for tasks like bricklaying, reducing construction time and labor costs.

Retail: The eBook explores automation in retail, from cashierless stores to warehouse robots for order fulfillment.

The British Perspective

The United Kingdom has played a significant role in the development and adoption of robotics and automation technologies. British companies, universities, and research institutions have contributed to the global advancements in these fields. "Robotics and Automation" recognizes the UK's contributions and encourages continued leadership in innovation and ethical considerations in the adoption of these technologies.

Challenges and Considerations

While robotics and automation offer remarkable advantages, they also present challenges, including concerns about job displacement, cybersecurity, and ethical considerations. The eBook emphasizes the importance of responsible adoption, ongoing training and upskilling, and cybersecurity measures to address these challenges.

A Vision for the Future

In conclusion, "Robotics and Automation" is not just a book; it is a roadmap to a future where human ingenuity and technological innovation intertwine to create a more efficient, productive, and interconnected world. It underscores that the robotics and automation revolution is not about replacing human workers but about augmenting their capabilities and improving their working conditions.

The eBook serves as a reminder that the future of industries lies in the synergy between humans and machines. It calls upon individuals, businesses, and governments to embrace robotics and automation as powerful tools for innovation and to navigate the evolving landscape with ethics, responsibility, and a commitment to positive change.

As we stand at the precipice of a new era in technology, "Robotics and Automation" inspires us to reimagine industries, reinvent processes, and reimpose human creativity in a world driven by efficiency and productivity. It is a testament to the boundless possibilities that arise when technology aligns with human ingenuity, paving the way for a brighter and more prosperous future for all.

CHAPTER 12

Virtual and Augmented Reality: A Journey into Immersive Worlds and Their Expansive Applications

In the realm of digital innovation, Virtual Reality (VR) and Augmented Reality (AR) stand as transformative technologies, beckoning us to step into immersive worlds and interact with digital information like never before. "Virtual and Augmented Reality" serves as an illuminating guide, unraveling the intricacies of VR and AR and their multifaceted applications beyond the realm of gaming. This exploration, steeped in British English, delves into the essence of VR and AR, their far-reaching impact, and their role in shaping the future of industries, education, healthcare, and more.

The Rise of Virtual and Augmented Reality

Virtual Reality and Augmented Reality are not just technologies; they are gateways to alternate realities and new dimensions of human experience. They represent the fusion of the digital and physical worlds, offering immersive experiences that transport us to places we could only dream of or enhance our understanding of the world around us.

Understanding Virtual and Augmented Reality

At their core, VR and AR technologies provide users with digitally enhanced experiences:

Virtual Reality (VR): VR creates entirely immersive digital environments that transport users to computer-generated worlds. Users typically wear headsets that cover their field of vision, providing a sense of presence within the virtual space.

Augmented Reality (AR): AR overlays digital information onto the real world, enhancing our perception of reality. AR experiences can be delivered through smartphones, tablets, or smart glasses, allowing users to interact with digital content while remaining aware of their physical surroundings.

Key Components and Technologies

"Virtual and Augmented Reality" explores the foundational elements that make these immersive experiences possible:

Hardware: It discusses the development of VR headsets and AR devices, such as the Oculus Rift, Microsoft HoloLens, and smartphone-based AR platforms.

Sensors: The eBook highlights the role of sensors like accelerometers and gyroscopes in tracking user movements, enabling natural interaction with virtual and augmented environments.

Content Creation: It delves into the creation of VR and AR content, including 3D modeling, animation, and real-time rendering.

Interactivity: The eBook emphasizes the importance of user interaction in VR and AR, such as hand tracking, haptic feedback, and spatial audio.

Applications Beyond Gaming

While VR and AR have roots in gaming, their applications extend far beyond entertainment:

Education: The eBook explores how VR and AR are revolutionizing education by providing immersive learning experiences, from virtual field trips to interactive anatomy lessons.

Healthcare: It discusses the use of VR in medical training, therapy, and pain management, as well as AR's role in assisting surgeons with real-time information during procedures.

Architecture and Design: It highlights how VR and AR are transforming the way architects and designers visualize and present their projects, allowing clients to explore virtual buildings before construction begins.

Retail: The eBook delves into how AR is enhancing the retail experience with virtual try-ons, interactive catalogs, and indoor navigation in stores.

Remote Work and Collaboration: It discusses how VR is enabling remote teams to collaborate in shared virtual spaces, improving communication and productivity.

Tourism: It showcases how VR is used to create virtual travel experiences, allowing users to explore destinations from the comfort of their homes.

The British Perspective

The United Kingdom has been actively involved in the development and adoption of VR and AR technologies. British companies, universities, and research institutions have contributed to the global advancement of these immersive technologies. "Virtual and Augmented Reality" recognizes the UK's contributions and encourages continued leadership in innovation and ethical considerations in the use of VR and AR.

Challenges and Considerations

While VR and AR offer exciting possibilities, they also raise challenges related to privacy, accessibility, and ethical use. The eBook underscores the importance of responsible development and application of these technologies, ensuring they benefit society as a whole.

A Vision for the Future

In conclusion, "Virtual and Augmented Reality" is not just a book; it is a portal to worlds uncharted and experiences unimagined. It underscores that VR and AR are not mere novelties but transformative tools that expand our horizons, enhance our understanding, and elevate our capabilities.

The eBook serves as a reminder that the future is immersive, a place where the boundaries between the digital and physical worlds blur, and where human experiences are enriched by technology. It calls upon individuals, businesses, and educators to embrace VR and AR as powerful tools for innovation and to navigate the evolving landscape with ethics, responsibility, and a commitment to unlocking the full potential of immersive technologies.

As we stand on the threshold of immersive realities, "Virtual and Augmented Reality" inspires us to step beyond the confines of the everyday and explore new dimensions of human experience, knowledge, and creativity. It is a testament to the boundless possibilities that arise when technology converges with human imagination, paving the way for a brighter and more enriched future for all.

CHAPTER 13

Quantum Computing: Unlocking the Potential of the Unthinkable

In the ever-evolving landscape of computing, quantum computing stands as a revolutionary force, challenging the boundaries of what is possible in the world of information processing. "Quantum Computing" serves as an illuminating guide, unveiling the extraordinary potential of quantum computers to solve complex problems that were once deemed insurmountable. In this exploration, steeped in British English, we delve into the essence of quantum computing, its transformative impact, and its role in shaping the future of science, technology, and innovation.

The Quantum Leap in Computing

Quantum computing represents a monumental shift in our approach to information processing. It harnesses the principles of quantum mechanics, which govern the behavior of subatomic particles, to perform computations that were previously beyond the capabilities of classical computers. It offers the promise of tackling complex problems in fields such as cryptography, materials science, optimization, and drug discovery with unprecedented efficiency.

Understanding Quantum Computing

At its core, quantum computing leverages the fundamental properties of quantum bits, or qubits, to perform calculations. Unlike classical bits, which can be either 0 or 1, qubits can exist in a superposition of states, allowing quantum computers to explore multiple solutions simultaneously. Additionally, qubits can be entangled, meaning the state of one qubit is intrinsically linked to the state of another, regardless of the physical distance between them.

Key Components and Technologies

"Quantum Computing" explores the foundational elements that make quantum computation possible:

Qubit Technologies: The eBook discusses various physical implementations of qubits, including superconducting circuits, trapped ions, and topological qubits, each with its advantages and challenges.

Quantum Gates: It highlights the role of quantum gates, analogous to classical logic gates, in manipulating qubits and performing quantum operations.

Quantum Algorithms: The eBook delves into quantum algorithms like Shor's algorithm, which has the potential to factor large numbers exponentially faster than classical computers, posing a threat to encryption systems.

Quantum Error Correction: Quantum computers are highly sensitive to errors, and the eBook explores error correction techniques to make quantum computations more reliable.

Applications Beyond Imagination

"Quantum Computing" delves into the extraordinary applications of quantum computing that are reshaping various fields:

Cryptography: Quantum computers have the potential to break widely used encryption methods, spurring the development of post-quantum cryptography to secure digital communications.

Drug Discovery: Quantum computing accelerates molecular simulations, enabling the discovery of new drugs and materials with remarkable precision.

Optimization: It discusses how quantum computing can optimize complex systems, from supply chains to financial portfolios, offering solutions in record time.

Climate Modeling: Quantum computers can simulate complex climate models with unprecedented accuracy, aiding in climate change research and mitigation strategies.

Machine Learning: The eBook explores the role of quantum computing in enhancing machine learning algorithms, enabling more efficient data analysis and pattern recognition.

The British Perspective

The United Kingdom has been actively involved in the development and application of quantum computing technologies. British researchers and institutions have contributed to the global advancements in this field. "Quantum Computing" recognizes the UK's contributions and encourages continued leadership in innovation and ethical considerations in the use of quantum computing.

Challenges and Considerations

While quantum computing holds immense promise, it also presents challenges, including technical hurdles, the need for quantum-safe encryption, and ethical considerations in areas like cryptography and privacy. The eBook emphasizes the importance of responsible development and application of quantum computing to ensure its benefits are realized while mitigating potential risks.

A Vision for the Future

In conclusion, "Quantum Computing" is not just a book; it is a window into a future where computation transcends the boundaries of classical limitations. It underscores that quantum computing is not merely an evolution of current technology but a revolution that promises to reshape industries, solve pressing global challenges, and push the boundaries of human knowledge.

The eBook serves as a reminder that quantum computing is a testament to human ingenuity and the relentless pursuit of understanding the universe. It calls upon individuals, businesses, and governments to embrace quantum computing as a powerful tool for innovation and to navigate the evolving landscape with ethics, responsibility, and a commitment to harnessing the full potential of quantum technologies.

As we stand at the threshold of a quantum-powered era, "Quantum Computing" inspires us to dream of a world where the uncharted becomes the routine, the complex becomes the solvable, and the impossible becomes the attainable. It is a testament to the boundless possibilities that arise when technology aligns with the deepest principles of the universe, paving the way for a brighter and more enlightened future for all.

CHAPTER 14

Ethical Tech: Navigating the Moral Compass of Technological Development

In our ever-advancing technological landscape, the compass of ethics guides us through the complex terrain of innovation. "Ethical Tech" serves as a guiding light, illuminating the profound importance of ethical considerations in technological development and offering insights on addressing emerging ethical challenges. In this exploration, steeped in British English, we delve into the essence of ethical tech, its pivotal role in shaping our digital future, and the strategies to ensure technology aligns with human values.

The Imperative of Ethical Tech

Technology has become an integral part of our lives, impacting everything from how we work to how we communicate and even how we perceive reality. With this ubiquity comes an ethical responsibility to ensure that technology, in all its forms, respects our values, safeguards our rights, and contributes positively to society.

Understanding Ethical Tech

Ethical tech encompasses a wide range of considerations and principles:

Privacy: It involves protecting individuals' privacy and data rights, addressing concerns related to data collection, surveillance, and data breaches.

Fairness: Ethical tech strives for fairness and equity in algorithmic decision-making, eliminating biases that can perpetuate discrimination.

Transparency: It emphasizes transparency in technology design, operation, and decision-making processes to build trust with users.

Accountability: Ethical tech holds developers, companies, and organizations accountable for the consequences of their technologies.

Security: It ensures that technology is secure against malicious attacks, protecting users' data and digital assets.

Key Considerations and Strategies

"Ethical Tech" explores key considerations and strategies to foster ethical development:

Ethical by Design: It advocates for incorporating ethical considerations from the outset of technology development, embedding values into the design process.

User-Centered Design: Prioritizing the user's well-being and rights in tech design ensures that technology serves human needs and aspirations.

Algorithmic Transparency: Understanding and revealing the inner workings of algorithms helps prevent biases and discriminatory outcomes.

Data Governance: Ethical tech calls for responsible data handling, including informed consent, data anonymization, and protection against data misuse.

Regulatory Frameworks: Governments and organizations should establish and enforce ethical guidelines and regulations to ensure compliance.

Emerging Ethical Challenges

"Ethical Tech" addresses several emerging ethical challenges:

AI and Bias: It discusses the challenge of bias in artificial intelligence and machine learning algorithms and the need for algorithmic fairness.

Surveillance Technology: Ethical concerns around surveillance technologies and their impact on privacy and civil liberties are explored.

synthetic media: The eBook delves into the ethical dilemmas posed by synthetic media and the potential for misinformation.

Autonomous Weapons: It discusses the ethical concerns surrounding the development and use of autonomous weapons systems.

The British Perspective

The United Kingdom has been actively engaged in promoting ethical tech practices. British companies, government bodies, and research institutions have contributed to the development of ethical guidelines and the advancement of responsible tech innovation. "Ethical Tech" recognizes the UK's contributions and encourages continued leadership in shaping global ethical tech standards.

Educational Initiatives

The eBook emphasizes the role of education in fostering ethical tech practices. It highlights the importance of raising awareness about digital ethics in schools, universities, and organizations to empower individuals with the knowledge and skills needed to navigate ethical challenges.

A Vision for the Future

In conclusion, "Ethical Tech" is not just a book; it is a call to action. It underscores that technology, while a powerful tool, is a reflection of human values and intentions. It reminds us that as we innovate, we must do so with the utmost respect for the well-being and dignity of individuals and society as a whole.

The eBook serves as a reminder that ethical tech is not a constraint but an opportunity—an opportunity to create a digital future that enhances our lives, protects our rights, and upholds our values. It calls upon individuals, businesses, policymakers, and technologists to champion ethical tech as a fundamental pillar of innovation and to navigate the evolving technological landscape with ethics, responsibility, and a commitment to the greater good.

As we stand at the crossroads of the digital age, "Ethical Tech" inspires us to forge a path that upholds the highest ethical standards, ensuring that technology truly becomes a force for positive change, equity, and a brighter future for all.

CHAPTER 15

Innovation Ecosystems: Nurturing Technological Progress through Collaboration and Creativity

In today's fast-paced world, innovation is the heartbeat of technological progress. "Innovation Ecosystems" serves as a comprehensive guide, shedding light on the pivotal role played by innovation hubs, startups, and collaborative environments in driving forward the frontiers of technology. Rooted in British English, this exploration unveils the essence of innovation ecosystems, their transformative impact, and their central role in shaping the future of industries, economies, and societies.

The Essence of Innovation Ecosystems

Innovation ecosystems are dynamic networks of organizations, individuals, and resources that converge to foster innovation, catalyze creativity, and accelerate the development of groundbreaking technologies. These ecosystems act as incubators, nurturing the seeds of innovation and providing fertile ground for them to grow and flourish.

Understanding Innovation Ecosystems

At their core, innovation ecosystems thrive on collaboration, knowledge sharing, and the convergence of diverse talents. They comprise several key components:

1. Innovation Hubs: These are physical or virtual spaces designed to encourage collaboration, experimentation, and idea exchange among startups, established companies, researchers, and entrepreneurs.
2. Startups: Small, agile, and innovative companies that are at the forefront of technological breakthroughs, often challenging established norms and industries.
3. Collaborative Environments: Platforms and spaces that encourage cross-disciplinary collaboration, often involving academia, research institutions, government bodies, and private enterprises.

Key Elements and Dynamics

"Innovation Ecosystems" delves into the essential elements that drive these ecosystems:

1. Open Innovation: It highlights the importance of open innovation models, where organizations collaborate with external partners to tap into a wider pool of ideas and expertise.
2. Investment and Funding: The eBook explores the role of venture capital, angel investors, and government grants in fueling the growth of startups and innovative projects.
3. Talent and Skills: It emphasizes the significance of attracting and retaining top talent in innovation hubs, creating a fertile ground for creative thinking and problem-solving.
4. Regulatory Support: The eBook discusses the importance of regulatory frameworks that encourage innovation while ensuring ethical and responsible development.

Applications Across Industries

"Innovation Ecosystems" showcases how these dynamic networks are making an impact across various sectors:

1. Technology: Innovation ecosystems have been instrumental in the development of cutting-edge technologies such as artificial intelligence, quantum computing, and biotechnology.
2. Healthcare: They are revolutionizing healthcare with startups working on personalized medicine, telemedicine, and medical device innovation.
3. Manufacturing: Collaborative environments are driving advancements in smart manufacturing and Industry 4.0, optimizing production processes.
4. Sustainability: Innovation ecosystems are fostering the creation of green technologies, renewable energy solutions, and sustainable practices across industries.
5. Finance: They are transforming the financial sector with financial technology startups, blockchain applications, and digital payment innovations.

The British Perspective

The United Kingdom has been actively involved in nurturing innovation ecosystems. British cities like London, Cambridge, and Manchester are renowned for their thriving innovation hubs, drawing talent and investment from around the world. "Innovation Ecosystems" recognizes the UK's contributions and encourages continued leadership in fostering creativity and innovation.

Challenges and Considerations

While innovation ecosystems offer immense potential, they also present challenges such as competition for talent, regulatory complexities, and the need for sustainable growth. The eBook underscores the importance of addressing these challenges through collaborative efforts and strategic planning.

A Vision for the Future

In conclusion, "Innovation Ecosystems" is more than a book; it is a testament to human ingenuity and the power of collective effort. It underscores that innovation ecosystems are not just catalysts for economic growth but engines of societal progress, shaping the way we live, work, and interact.

The eBook serves as a reminder that innovation ecosystems are about nurturing creativity, celebrating diversity, and fostering an environment where ideas can flourish. It calls upon individuals, businesses, policymakers, and educators to champion these ecosystems as essential pillars of progress and to navigate the evolving landscape with a commitment to fostering innovation, collaboration, and a brighter future for all.

As we stand at the crossroads of innovation, "Innovation Ecosystems" inspires us to embrace collaboration, celebrate diversity, and unlock the potential of human creativity to address the challenges of our time and create a more prosperous and sustainable world for generations to come.

www.ingramcontent.com/pod-product-compliance
Lightning Source LLC
Chambersburg PA
CBHW080942260726
48661CB00010B/4050